WHAT ARE AMPHIBIANS?, WHAT & WHY: IST GRADE SCIENCE SERIES

SPEEDY
PUBLISHING

Speedy Publishing LLC
40 E. Main St. #1156
Newark, DE 19711
www.speedypublishing.com

Copyright 2015

The word amphibian means two-lives. Amphibians spend their lives in the water and on land.

Amphibians usually have soft, moist skin that is protected by a slippery layer of mucus.

All amphibians
are
vertebrates,
which means
they have
a backbone
or spine.

Amphibians are cold-blooded, which means that they are the same temperature as the air or water around them.

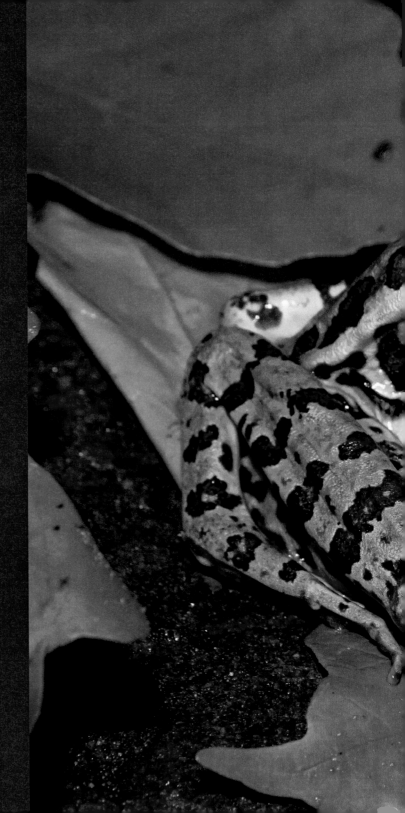

Amphibians have adapted to live in a number of different habitats including streams, forests, meadows, bogs, swamps, ponds, rainforests, and lakes.

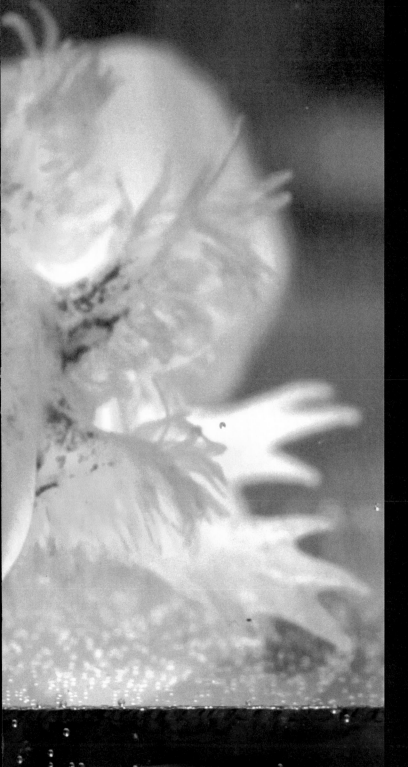

An Amphibian's skin absorbs air and water. This makes them very sensitive to air and water pollution.

Amphibians are herbivores as larvae and carnivores as adults. Once they turned into adults, they will eat any animal that is small enough to be swallowed whole.

Amphibians have been around a long time. The earliest known amphibian fossil dates from 368 million years ago.

Made in the USA
Middletown, DE
25 May 2016